ULTIMATE SUPERCARS

CHEVROLET CAMARO ZL1

By Tamra B. Orr

Kaleidoscope
Minneapolis, MN

The Quest for Discovery Never Ends

This edition first published in 2023 by Kaleidoscope Publishing, Inc.

No part of this publication may be reproduced in whole or in part without written permission of the publisher.

For information regarding permission, write to
Kaleidoscope Publishing, Inc.
6012 Blue Circle Drive
Minnetonka, MN 55343

Library of Congress Control Number
2022937983

ISBN
978-1-64519-609-9 (library bound)
978-1-64519-679-2 (ebook)

Text copyright © 2023 by Kaleidoscope Publishing, Inc. All-Star Sports, Bigfoot Books, and associated logos are trademarks and/or registered trademarks of Kaleidoscope Publishing, Inc.

Printed in the United States of America.

FIND ME IF YOU CAN!

Bigfoot lurks within one of the images in this book. It's up to you to find him!

TABLE OF CONTENTS

Chapter 1: The Chevy Beast 4

Chapter 2: From Panther to Bumblebee 12

Chapter 3: A Chevy Dragon 18

Chapter 4: Saying Farewell 24

Beyond the Book 28
Research Ninja 29
Further Resources 30
Glossary 31
Index 32
Photo Credits 32
About the Author 32

Chapter 1
The Chevy Beast

A total beast.

A road maniac.

A demon at the drag strip.

The king of monster **muscle cars**.

The Chevy Camaro ZL1 has been called these names. Why? Because it is so powerful and fast! With a 650-horsepower supercharged V-8 engine, it eats up the miles with almost no effort. *Car and Driver* says that the 2022 model "makes its driver feel like a king and lords over its fiercest rivals."

Driving the ZL1 down the road is exciting. When its engine roars, one reviewer says, "It leaves the sound of thunder echoing in its wake."

Chevy Camaros have been around for over 50 years! Chevy introduced the ZL1 model in 2017. Each year, they make a few changes. They improve the handling. They add safety equipment. They make the car look and move faster.

The power from the ZL1's engine is sent to all four of its wheels at once. This means one push on the gas pedal, and the car goes from 0 to 100 miles per hour (160 km/h) in 7.4 seconds! Although the engine is powerful, Chevy has worked hard to keep road noise and vibration down. It's a smooth ride.

PARTS OF A
CAMARO ZL1

carbon-fiber gas cap

19-inch (48 cm) forged aluminum wheels

6-piston front and 4-piston rear brake calipers

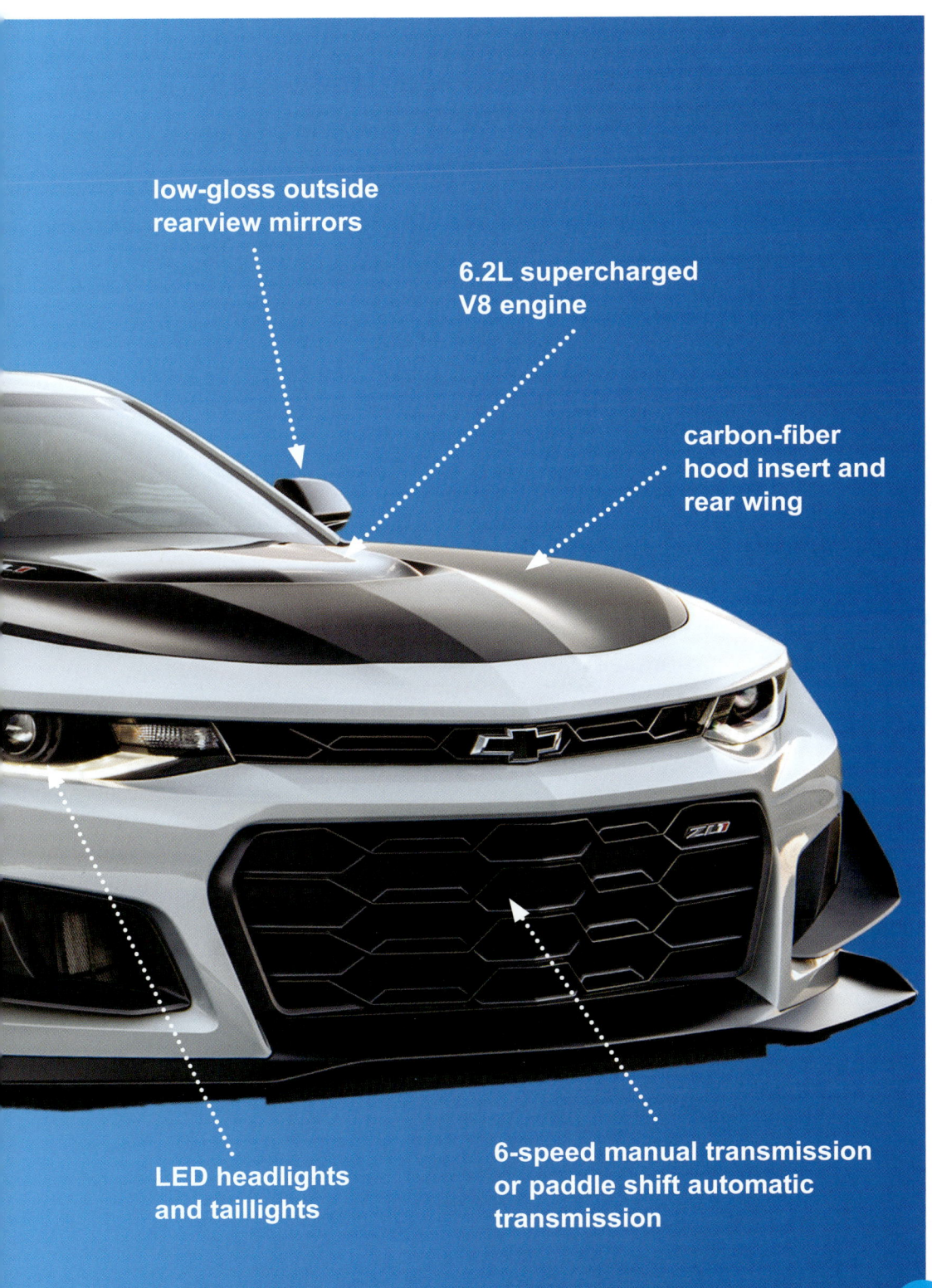

Since Chevy first began making Camaros, the cars have gotten more high-tech. The new ZL1 has a *magnetic ride control system*. The car can sense changes in the road ahead. In **milliseconds**, it can adjust the car's **shock absorbers** to make the ride smoother.

The high-tech options do not stop there. The newest models offer a tire pressure monitoring system. Drivers will know the second a tire needs more air! It has a rear camera to make parking easier. The lane change alert lets drivers know if it is safe to change lanes.

FUN FACT

The first Chevy Camaro ZL1s were only available in five colors: Hugger Orange, Cortez Silver, Code-51 Dusk Blue, LeMans Blue, and Fathom Green.

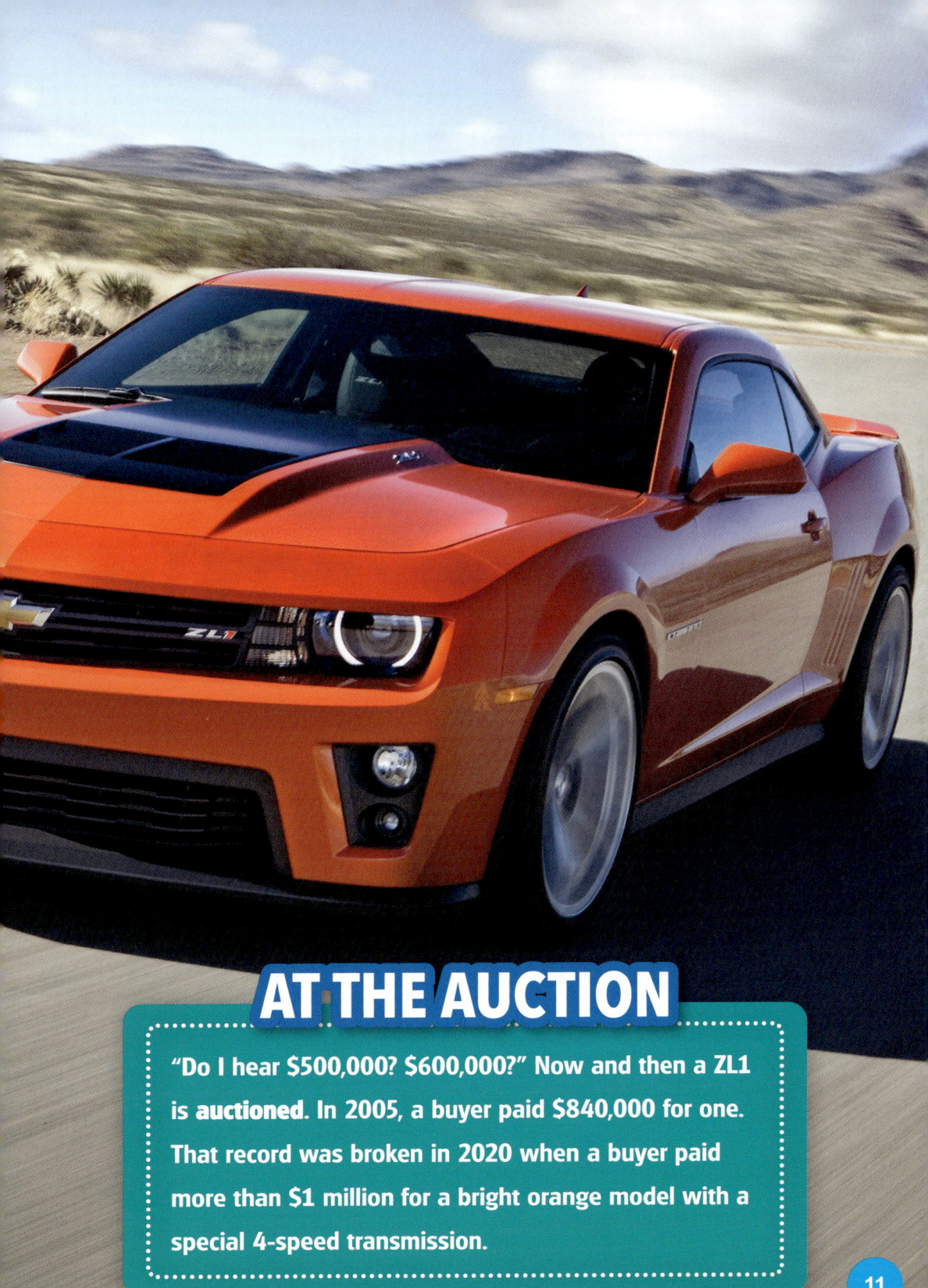

AT THE AUCTION

"Do I hear $500,000? $600,000?" Now and then a ZL1 is **auctioned**. In 2005, a buyer paid $840,000 for one. That record was broken in 2020 when a buyer paid more than $1 million for a bright orange model with a special 4-speed transmission.

Chapter 2
From Panther to Bumblebee

Chevrolet first began working on the Camaro model in the mid-1960s. They planned to call it the Panther. That did not last long. Chevy was already producing the Chevelle, the Corvette, and the Corvair. It needed another name that started with "C." The Camaro was born!

Some say the name was based on the Spanish word for "little friend." Others say they made it up. When the first Camaros were made, the Ford Mustang was one of the most popular cars. One General Motors manager stated Camaro meant "a small, vicious animal that eats Mustangs."

The first Camaro went on sale in September 1966 for $2,466!

There have been six generations of Camaros. The ZL1 was the first Chevrolet model to offer a 10-speed automatic transmission to owners.

In 2007, a model from the second generation grabbed the attention of movie watchers everywhere. This beat-up, rusty yellow 1977 Camaro was a star of the first *Transformers* movie. *Bumblebee* could only talk by playing songs on its radio. The transforming Camaro appeared in all three of the movie's sequels and got its own movie in 2018.

Each movie updated the Camaro to make it look leaner and meaner.

WHERE THE CAMARO ZL1 IS MADE

Lansing, Michigan

AMERICAN MADE

Chevy is a part of General Motors. There are 122 General Motors facilities. Camaros are made at the Lansing Grand River Assembly in Lansing, Michigan. It is one of their newest plants.

FUN FACT

In 2013, the police patrol cars in Dubai in the United Arab Emirates were Camaro SS coupes, thanks to the model's speed.

In the beginning of 2001, Camaro sales were falling. But, thanks to the car's role in *Transformers*, people were asking Chevrolet to make a new model. They did. Sales soared. The car became even more popular when the car company offered its first convertible model.

Today, the ZL1 is considered one of the most powerful cars on the market. There are no plans for a seventh generation, though. Chevrolet announced in 2021 that it was going to stop making the model. It plans to only produce electric cars by 2035. Time will tell if there will one day be an all-electric Camaro.

Chapter 3
A Chevy Dragon

Today's Camaro ZL1 rumbles like a hungry dragon, thanks to its dual-mode exhaust. Found on the back, this system controls the volume of the exhaust. Step on the gas, and the dragon roars! The long, thin LED headlights and taillights look like dragon eyes glowing in the darkness.

The Camaro's engine is like a dragon that breathes fire. As the car races down the road, the engine can get hot, so the ZL1 has a grille with huge stacked air intake slots. Here outside air keeps the engine and radiator nice and cool.

THE CAMARO ZL1
IN DETAIL

COST: $63,000 basic model

Height: 4.4 feet (1.3 m)

Width: 6.25 feet (1.9 m)

LENGTH: 15.85 feet (4.83 m)

WEIGHT: 3,907 pounds (1,772 kg)

TOP SPEED: 202 miles per hour (325 km/h)

TIME FROM 0 to 60 miles per hour (96 km/h): 3.5 seconds

Slide inside a ZL1 and it is like time traveling into the future. It has so many high-tech features. It even comes in new colors, including metallic shades of gray and blue.

The Lane Keep Assist feature gently pulls the steering wheel back to your lane if you drift into the other lane. Park Assist uses cameras showing images of the car from the front, sides, and rear. The cameras also let you know if a child, bicycle, or animal is too close. The Intellibeam Assist controls your headlights. They are bright until a car comes from the other direction. Then they instantly dim until the car is gone.

THE LE MODEL

Want the ZL1 to be even more high-performance? Many owners add in the 1LE package for $7,500. It adds a carbon-fiber hood insert and rear **spoiler**. Throw in some wide 19-inch (48-cm) aluminum wheels and a 6-speed transmission, and you are ready for road racing.

FUN FACT

The new ZL1 has five driving **modes**: Normal, Stealth, Sport, Tour, and Track. Changing modes even changes the engine's roar.

Chapter 4
Saying Farewell

When Chevy announced they would no longer be making the Camaro after 2024, people were sad. The car has been around for over 50 years. Chevrolet has stated, however, that they are making a special farewell package for the last model. They will only make 2,000 of them.

What will this special model have? No one is sure, but there have been rumors! It may have an extra 170 horsepower and new accents like racing stripes. To honor *Bumblebee*, these cars will all be bright yellow.

For more than half a century, Camaro has been one of the cars people dreamed of owning and driving. They saw it in movies. They watched it win international races. They read about its powerful engine and its nicknames of beast and monster. Saying goodbye to this line of muscle cars is hard, but the world of car production is heading towards all-electric.

FUN FACT
The ZL1 can seat four, as long as the people sitting in the back are not too long-legged.

All-electric cars are better for the planet and people. Air pollution will improve. It is the right way for car manufacturers to go. But there is no question about it. This is one dragon that will be missed.

BANNING THE COPPER

What do Camaros have to do with salmon? As of 2021, the ZL1, SS, or 1LE models are illegal in California and Washington State. Their brake pads have high amounts of copper. As the cars are driven, the copper leaks. Over time, it reaches waterways and can hurt salmon.

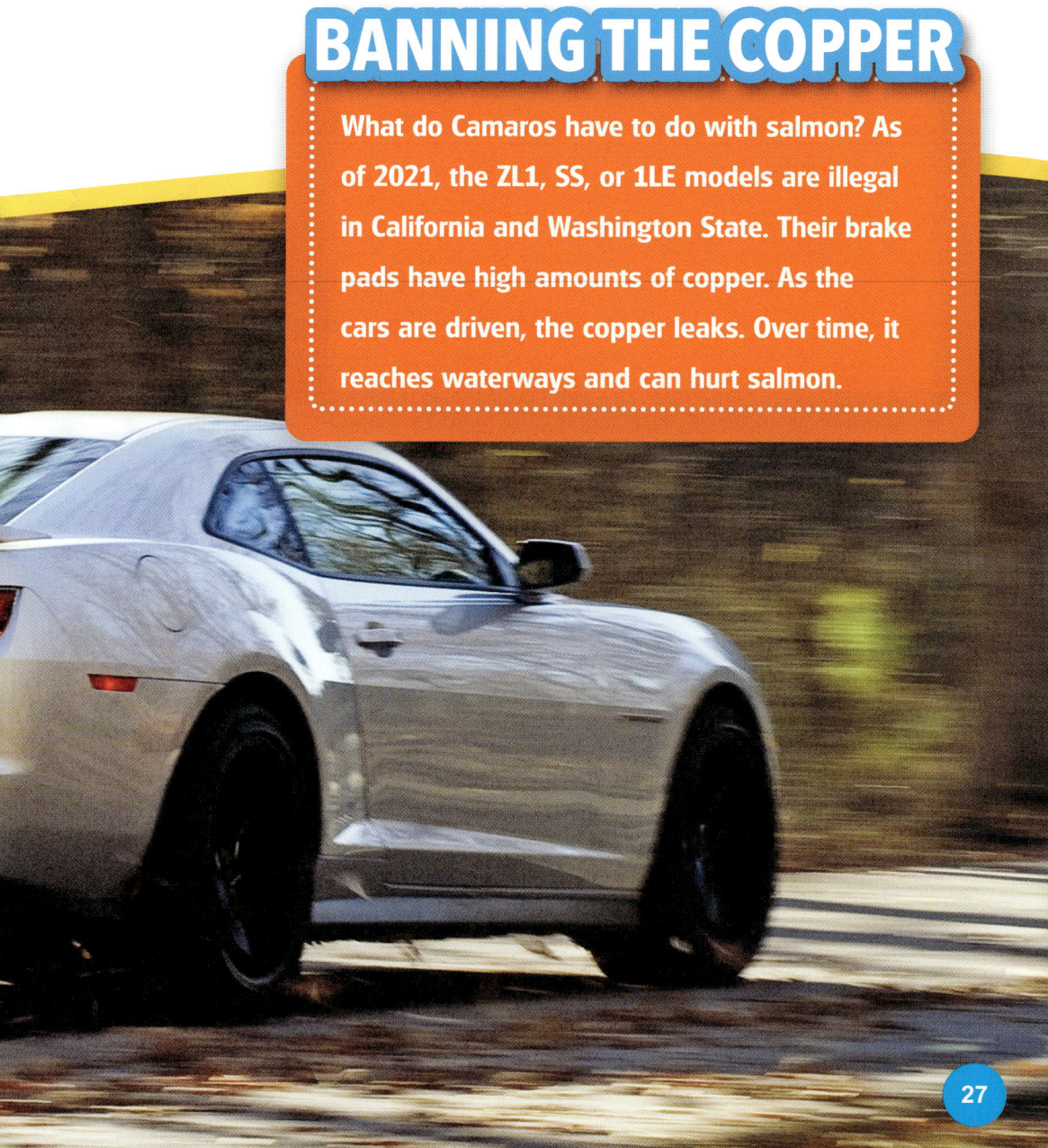

BEYOND THE BOOK

After reading the book, it's time to think about what you learned. Try the following exercises to jump-start your ideas.

THINK

THAT'S NEWS TO ME. The last model year of the ZL1 is 2024. How might news sources be able to fill in more details about what the last model will include? What information could you find in news articles? Where could you go to find those sources?

CREATE

PRIMARY SOURCE. A primary source is an original document, photograph, or interview. Make a list of primary sources you might be able to find about the ZL1. What new information might you learn from these sources?

SHARE

WHAT'S YOUR OPINION? This book claims that the ZL1 is considered one of the most powerful cars on the market. Do you agree or disagree with this position? Use evidence from the text to support your answer. Share your position and evidence with a friend. Does your friend agree with you?

GROW

REAL-LIFE RESEARCH. What places could you visit to learn more about the Camaro ZL1? What other things could you learn while you were there?

RESEARCH NINJA

Visit **www.ninjaresearcher.com/6099** to learn how to take your research skills and book report writing to the next level!

RESEARCH

DIGITAL LITERACY TOOLS

SEARCH LIKE A PRO
Learn about how to use search engines to find useful websites.

FACT OR FAKE?
Discover how you can tell a trusted website from an untrustworthy resource.

TEXT DETECTIVE
Explore how to zero in on the information you need most.

SHOW YOUR WORK
Research responsibly— learn how to cite sources.

WRITE

GET TO THE POINT
Learn how to express your main ideas.

PLAN OF ATTACK
Learn prewriting exercises and create an outline.

DOWNLOADABLE REPORT FORMS

Further Resources

BOOKS

Cruz, Calvin. *Chevrolet Corvette Z06*. Bellwether Media: Minnetonka, MN, 2014.

Hayes, Jaxon. *Chevrolet Camaro ZL1 1LE vs. Dodge Challenger SRT Hellcat Redeye*. Gray Duck Creative Works: Minneapolis, MN, 2021.

Rusick, Jessica. *Chevrolet Camaro*. Big Buddy Books: Eagle, Idaho, 2020.

WEBSITES

FACTSURFER

Factsurfer.com gives you a safe, fun way to find more information.

1. Go to www.factsurfer.com.
2. Enter "Chevy Camaro ZL1" into the search box and click.
3. Select your book cover to see a list of related websites.

Glossary

auctioned: when goods are sold at a public sale, to the highest bidder.

calipers: a vehicle brake with two hinges.

convertible: a car with a folding top.

coupe: a car with a fixed roof and two doors.

millisecond: one-thousandth of a second.

modes: different driving options based on conditions.

muscle cars: high-performance cars often used for racing.

shock absorbers: a device for reducing jolts and vibrations in a vehicle.

spoiler: a device on the front or rear of a vehicle to prevent it from being lifted off of the road when traveling at high speeds.

Index

Bumblebee, 12, 14, 24
camera, 10, 17, 22
Chevrolet, 6, 14, 16, 17, 24
engine, 4, 7, 9, 18, 26
horsepower, 4, 24
LED, 9, 18
model, 4, 6, 10, 11, 12, 14, 16,
    17, 20, 22, 24, 27
Transformers, 14, 16
transmission, 9, 11, 14, 22
wheel, 7, 8, 22

PHOTO CREDITS

The images in this book are reproduced through: Veyron Photo/Shutterstock 14. All other images courtesy of Chevrolet Pressroom/General Motors (Jesse Lynn Walker 4-5, 6-7, 8-9, 18-19). **Cover:** Courtesy of Chevrolet Pressroom/General Motors, Ilyshev Dmitry/Shutterstock (background).

About the Author

Tamra B. Orr is a full-time author living in the Pacific Northwest with her family. She attended Ball State University before moving cross-country. Orr has written more than 750 books for readers of all ages and says it is the best job in the world because she is always learning something new.